ε Epsilon

Fractions

Tests

1-888-854-MATH (6284) - mathusee.com
sales@mathusee.com

Epsilon Tests: Fractions

©2012 Math-U-See, Inc.
Published and distributed by Demme Learning.

mathusee.com

1-888-854-6284 or +1 717-283-1448 | demmelearning.com
Lancaster, Pennsylvania USA

ISBN 978-1-60826-076-8
Revision Code 1118-D

Printed in the United States of America by The P.A. Hutchison Company
2 3 4 5 6 7 8 9 10

For information regarding CPSIA on this printed material call: 1-888-854-6284
and provide reference #1118-11242020

LESSON TEST

Build the problem and then write the correct solution.

1. **Step 1.** Select 12 blocks.

    Step 2. Divide into 2 equal parts.

    Step 3. Count 1 of them.

    —— of ___ is ___

2. **Step 1.** Select 16 blocks.

    Step 2. Divide into 4 equal parts.

    Step 3. Count 3 of them.

    —— of ___ is ___

Read the problem, build it, and write the answer in the blank.

3. One third of fifteen is _____.

4. Two fifths of twenty is _____.

Solve.

5. $\frac{1}{8}$ of 16 = _____

6. $\frac{3}{4}$ of 8 = _____

7. $\frac{1}{2}$ of 10 = _____

8. $\frac{3}{6}$ of 36 = _____

9. $\frac{2}{7}$ of 49 = _____

10. $\frac{4}{5}$ of 40 = _____

Find the perimeter of each rectangle.

11.

5"

8"

P = _____

12.

12'

24'

P = _____

13.

32"

55"

P = _____

14. One sixth of the people in the room have blue eyes. If there are 18 people in the room, how many have blue eyes?

15. Kimberly bought a book that cost $14.58 and a box of paints for $6.75. How much money did she spend?

16. Kimberly had $25 before she went shopping (#15). How much money did she have after buying her book and paints?

17. Allison went on a two-week vacation. If it rained $\frac{2}{7}$ of the days she was on vacation, how many days did it rain?

18. Raleigh built a yard for his dog. The yard was a rectangle that measured 12 feet by 15 feet. How many feet of fence did Raleigh need to buy?

Write the correct numerator and denominator in both symbols and words.

1. —

2. —

3. —

Build the fraction and then draw it in the square. Say it and write it with numerals or words.

4. —

four fifths

5. —

one third

6. $\frac{1}{2}$

Solve.

7. $\frac{2}{5}$ of 25 = _____

8. $\frac{1}{10}$ of 80 = _____

9. $\frac{3}{7}$ of 42 = _____

Find the perimeter of each square or rectangle.

10.

32 yd

81 yd

P = _____

11.

28'

28'

P = _____

12. Thirty-five people tried out for the play. Three fifths of them were chosen to be in the play. How many people were chosen to be in the play?

13. What is the perimeter of a square patio that measures 10 feet on a side?

14. A bench was built along $\frac{1}{5}$ of the perimeter of the patio in #13. How long is the bench?

15. Chance had $16.45 in his wallet. He lost $5.30. How much money does Chance have left in his wallet?

LESSON TEST

Build, shade, and then write your answer in words and symbols.

1.

$$\square \quad + \quad \square \quad = \quad \square$$

___ + ___ = ___

one sixth plus four sixths equals _____

2.

$$\square \quad - \quad \square \quad = \quad \square$$

___ - ___ = ___

three fourths minus one fourth equals _____

Add or subtract.

3. $\dfrac{3}{5} + \dfrac{2}{5} =$ _____

4. $\dfrac{4}{6} + \dfrac{1}{6} =$ _____

5. $\dfrac{2}{10} + \dfrac{6}{10} =$ _____

6. $\dfrac{5}{8} - \dfrac{3}{8} =$ _____

7. $\dfrac{8}{9} - \dfrac{2}{9} =$ _____

8. $\dfrac{2}{3} - \dfrac{1}{3} =$ _____

Solve.

9. $\dfrac{1}{8}$ of 24 = _____

10. $\dfrac{4}{5}$ of 10 = _____ .

11. $\frac{4}{10}$ of 20 = _____

Find the perimeter of each shape.

12.   P = _____

13.
    8 yd

    12 yd   P = _____

14.
    16"

    16"   P = _____

15. Ben ate $\frac{1}{6}$ of an apple pie and $\frac{2}{6}$ of a blueberry pie. How much pie did he eat in all?

16. Katherine has finished $\frac{2}{3}$ of the job her mother gave her. What part of the job does she have left to finish?

17. Elizabeth plans to spend $\frac{3}{4}$ of the year at school in New Hampshire. For how many months will she be in New Hampshire?

18. Ruth and Clara had a box of one dozen doughnuts. Ruth ate $\frac{1}{6}$ of them, and Clara ate $\frac{2}{6}$ of them. What part of the doughnuts have been eaten? How many doughnuts have been eaten?

19. What part of the doughnuts in #18 is left to eat? How many doughnuts is that?

20. Caleb received $15.50, $5.25, and $10.00 as gifts. How much money did he receive in all?

LESSON TEST

Build the equivalent fractions. Draw lines and fill in the blanks to show what you have built. Write the fractions with numbers and with words.

1.

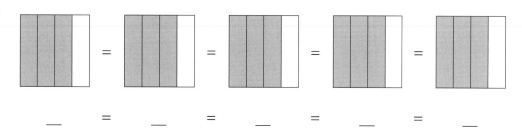

___ = ___ = ___ = ___ = ___

_____ _____ _____ _____ _____

Fill in the missing numbers to make equivalent fractions.

2. $\frac{1}{5} = \underline{\qquad} = \frac{}{15} = \frac{4}{}$

3. $\frac{5}{6} = \underline{\qquad} = \frac{15}{} = \frac{}{24}$

4. $\frac{3}{7} = \underline{\qquad} = \underline{\qquad} = \frac{12}{28}$

5. $\frac{1}{2} = \underline{\qquad} = \frac{3}{} = \frac{}{8}$

Add or subtract.

6. $\frac{4}{5} - \frac{1}{5} = \underline{\qquad}$

7. $\frac{5}{8} + \frac{2}{8} = \underline{\qquad}$

8. $\frac{5}{9} - \frac{4}{9} = \underline{\qquad}$

Solve.

9. $\frac{3}{8}$ of 48 = $\underline{\qquad}$

10. $\frac{1}{6}$ of 18 = $\underline{\qquad}$

11. $\frac{2}{3}$ of 27 = $\underline{\qquad}$

Multiply.

12.   2 2
       × 3 1

13.   5 4
       × 1 1

14.   3 2 2
       × 2 2

15. Lauren has colored one half of the pages in her coloring book. How many fourths has she colored?

16. If Lauren's book (#15) has 20 pages, how many pages has she colored?

17. If Rebekah is 13 years old, for how many months has she lived?

18. Each side of a square measures 11 feet. What is the perimeter of the square?

19. Annika read $\frac{2}{7}$ of her book before lunch and $\frac{3}{7}$ of it after lunch. What part of her book does she have left to read?

20. Annika's book has 56 pages (#19). How many pages does she have left to read?

Use the overlays to find equivalent fractions and then add or subtract.

1. $\dfrac{3}{6} + \dfrac{2}{5} =$ _____

2. $\dfrac{2}{4} + \dfrac{1}{3} =$ _____

3. $\dfrac{1}{5} + \dfrac{1}{6} =$ _____

4. $\dfrac{1}{4} - \dfrac{1}{5} =$ _____

5. $\dfrac{5}{6} - \dfrac{2}{5} =$ _____

6. $\dfrac{1}{2} - \dfrac{1}{4} =$ _____

Fill in the missing numbers in the numerators or denominators to make equivalent fractions.

7. $\dfrac{1}{7} =$ _____ $= \dfrac{}{21} = \dfrac{4}{}$

8. $\dfrac{2}{3} =$ _____ $=$ _____ $= \dfrac{8}{12}$

Solve.

9. $\dfrac{5}{9}$ of 36 = _____

10. $\dfrac{3}{10}$ of 90 = _____

11. $\dfrac{4}{6}$ of 48 = _____

Round to the nearest ten.

12. $85 \rightarrow$ _____

13. $13 \rightarrow$ _____

Round to the nearest hundred.

14. $250 \rightarrow$ _____

15. $175 \rightarrow$ _____

16. There were 18 soliders on duty. One sixth of the soldiers rode in tanks, and one third of them rode in trucks. The rest of them had to walk. What part of the soldiers was able to ride in a vehicle? (Use the overlays.)

17. What part of the soldiers in #16 had to walk?

18. Zarah bought $\frac{2}{3}$ of a pound of nuts. She used $\frac{1}{2}$ of a pound of nuts in the cake she baked. What part of a pound of nuts is left over? (Use the overlays.)

19. Sharon makes and sells bread. If each loaf takes three cups of flour, how much flour will she need for 122 loaves?

20. The sides of a triangle are 7 feet, 24 feet, and 25 feet. What is the perimeter of the triangle?

Add or subtract using the Rule of Four.

1. $\dfrac{1}{3} + \dfrac{2}{5} = $ _____

2. $\dfrac{3}{4} + \dfrac{1}{6} = $ _____

3. $\dfrac{3}{7} + \dfrac{1}{10} = $ _____

4. $\dfrac{1}{2} - \dfrac{1}{6} = $ _____

5. $\dfrac{5}{8} - \dfrac{1}{3} = $ _____

6. $\dfrac{5}{6} - \dfrac{3}{4} = $ _____

Fill in the missing numbers in the numerators or denominators to make equivalent fractions.

7. $\dfrac{7}{8} = $ _____ $= \dfrac{}{24} = \dfrac{28}{}$

8. $\dfrac{1}{3} = $ _____ $= $ _____ $= \dfrac{}{12}$

Solve.

9. $\dfrac{1}{10}$ of 50 = _____

10. $\dfrac{3}{4}$ of 16 = _____

11. $\dfrac{1}{2}$ of 20 = _____

Estimate and then multiply to find the exact answer.

12. $\begin{array}{r} 26 \\ \times\, 33 \\ \hline \end{array}$ →

13. $\begin{array}{r} 75 \\ \times\, 83 \\ \hline \end{array}$ →

14. $\begin{array}{r} 61 \\ \times\, 17 \\ \hline \end{array}$ →

15. Sixteen boxes of books arrived in the mail. If each box held one dozen books, how many books did I receive?

16. One half of Hannibal's elephants died as they struggled over the Alps. Another one fifth did not survive the march through Italy. What part of his elephants did Hannibal lose?

17. Mom used $\frac{4}{9}$ of the apples to make applesauce and $\frac{3}{6}$ of of them to make a pie. What part of the apples has been used?

18. Gavin saw $\frac{3}{4}$ of a pizza on the counter. By the time he was finished eating, there was only $\frac{1}{8}$ of a pizza left. What part of a pizza did Gavin eat?

19. Round 472 to the nearest hundred.

20. Vontoria took $\frac{7}{10}$ of an hour to do her math today. How many minutes did she spend on math today?

Use the Rule of Four to compare the fractions and write the correct symbol in the oval.

1. $\dfrac{1}{4}$ ◯ $\dfrac{3}{7}$

2. $\dfrac{3}{8}$ ◯ $\dfrac{1}{2}$

3. $\dfrac{4}{5}$ ◯ $\dfrac{2}{9}$

4. $\dfrac{6}{11}$ ◯ $\dfrac{2}{3}$

5. $\dfrac{5}{9}$ ◯ $\dfrac{6}{7}$

6. $\dfrac{3}{4}$ ◯ $\dfrac{6}{8}$

Add or subtract.

7. $\dfrac{1}{8} + \dfrac{4}{9} = $ _____

8. $\dfrac{2}{3} - \dfrac{1}{5} = $ _____

9. $\dfrac{4}{5} + \dfrac{1}{6} = $ _____

Fill in the missing numbers in the numerators and denominators to make equivalent fractions.

10. $\dfrac{2}{5} = $ _____ $=$ _____ $=$ _____

Divide. Include a fraction in the answer if the problem does not divide evenly.

11. $2\overline{)19}$

12. $6\overline{)51}$

13. $5\overline{)39}$

Estimate and then multiply to find the exact answer.

14. $\begin{array}{r} 39 \\ \times 24 \\ \hline \end{array}$ →

15. $\begin{array}{r} 72 \\ \times 15 \\ \hline \end{array}$ →

16. $\begin{array}{r} 68 \\ \times 43 \\ \hline \end{array}$ →

17. Five eighths of the trees in my yard are maples. If there are 16 trees in my yard, how many are maple trees?

18. Christa ate $\frac{1}{5}$ of the chocolates in the box, and Douglas ate $\frac{3}{10}$ of them. Write a comparison showing who ate more chocolates.

19. If there were 20 chocolates in the box in #18, how many chocolates did each person eat? Write another comparison using the actual number of chocolates eaten. Does it agree with the comparison you wrote for #18?

20. Yesterday we got $\frac{2}{10}$ of an inch of rain. Today we had $\frac{8}{10}$ of an inch of rain. How much more rain did we get today?

LESSON TEST

Use whichever method you prefer to add the fractions.

1. $\dfrac{1}{2} + \dfrac{3}{7} + \dfrac{1}{3} =$ ———

2. $\dfrac{4}{6} + \dfrac{1}{4} + \dfrac{2}{3} =$ ———

3. $\dfrac{3}{4} + \dfrac{3}{8} + \dfrac{1}{2} =$ ———

4. $\dfrac{3}{4} + \dfrac{1}{6} + \dfrac{1}{5} =$ ———

5. $\dfrac{2}{8} + \dfrac{1}{2} + \dfrac{2}{5} =$ ———

6. $\dfrac{1}{2} + \dfrac{2}{3} + \dfrac{5}{6} =$ ———

Use the Rule of Four to compare the fractions and write the correct symbol in the oval.

7. $\dfrac{2}{4} \bigcirc \dfrac{4}{8}$

8. $\dfrac{3}{7} \bigcirc \dfrac{3}{8}$

9. $\dfrac{2}{5} \bigcirc \dfrac{3}{4}$

Fill in the missing numbers in the numerators or denominators to make equivalent fractions.

10. $\dfrac{3}{9} =$ ——— $=$ ——— $= \dfrac{}{36}$

11. ——— $= \dfrac{4}{6} =$ ——— $= \dfrac{8}{}$

Divide. Include a fraction in the answer if the problem does not divide evenly.

12. 3 | 2 9 8   13. 7 | 1 5 6

14. 4 | 4 6 5

15. Brayden planted tomatoes in his garden. One fourth of the plants were eaten by pests. One half of them died because Brayden forgot to water them, and one eighth of the plants were killed when the dog dug a hole in the garden. What part of Brayden's tomato plants died or were destroyed?

16. What part of Brayden's plants survived? (#15) If he started with 32 plants, how many plants are left?

17. Each of Brayden's surviving plants (#16) produced 65 tomatoes during the summer. How many tomatoes did the plants produce altogether?

18. Mr. Knowles divided $217 evenly among his five children. How many dollars did each child receive? (Include a fraction in the answer if the problem does not divide evenly.)

19. What is 61 rounded to the nearest ten?

20. What is 829 rounded to the nearest hundred?

UNIT TEST Lessons 1-8

Solve.

1. $\dfrac{2}{3}$ of 12 = _____

2. $\dfrac{3}{5}$ of 10 = _____

3. $\dfrac{3}{4}$ of 24 = _____

4. $\dfrac{1}{3}$ of 15 = _____

5. $\dfrac{4}{5}$ of 25 = _____

6. $\dfrac{5}{7}$ of 14 = _____

Fill in the missing numbers in the numerators or denominators to make equivalent fractions.

7. $\dfrac{2}{6}$ = _____ = _____ = $\dfrac{}{24}$

8. _____ = $\dfrac{10}{16}$ = _____ = $\dfrac{20}{}$

Add or subtract using the Rule of Four when needed.

9. $\dfrac{1}{3} + \dfrac{1}{3}$ = _____

10. $\dfrac{1}{4} + \dfrac{3}{6}$ = _____

11. $\dfrac{1}{5} + \dfrac{5}{6}$ = _____

12. $\dfrac{2}{3} - \dfrac{1}{5}$ = _____

13. $\dfrac{7}{10} - \dfrac{3}{10} =$ _____

14. $\dfrac{4}{7} - \dfrac{3}{8} =$ _____

Use the Rule of Four to compare the fractions. Write the correct symbol in the oval.

15. $\dfrac{3}{5}$ ◯ $\dfrac{5}{7}$

16. $\dfrac{3}{6}$ ◯ $\dfrac{4}{8}$

17. $\dfrac{2}{3}$ ◯ $\dfrac{4}{8}$

Use whichever method you prefer to add the fractions.

18. $\dfrac{3}{8} + \dfrac{1}{10} + \dfrac{2}{5} =$ _____

19. $\dfrac{3}{4} + \dfrac{7}{8} + \dfrac{1}{2} =$ _____

Find the perimeter of each shape.

20.
28'
17'
16'

P = _____

21.
23 yd
46 yd

P = _____

22.
52"
52"

P = _____

Round to the nearest ten.

23. 91 → _____

24. 45 → _____

Round to the nearest hundred.

25. 604 → _____

26. 517 → _____

Divide. Include a fraction in the answer if the problem does not divide evenly.

27. $5\overline{)358}$

28. $8\overline{)541}$

29. $6\overline{)189}$

Multiply.

30. $\begin{array}{r} 6\,7 \\ \times\,1\,8 \\ \hline \end{array}$

31. $\begin{array}{r} 3\,4 \\ \times\,2\,6 \\ \hline \end{array}$

32. $\begin{array}{r} 2\,2\,4 \\ \times\,2\,2 \\ \hline \end{array}$

33. Sophia used $\frac{2}{5}$ of her birthday money for school supplies and $\frac{1}{3}$ of it for gifts. What part of her birthday money has she spent so far?

34. During our last snow storm, Drumore received $\frac{5}{6}$ of a foot, and Quarryville received $\frac{2}{3}$ of a foot. Tell which town received more snow. Then find the difference in the amounts of snowfall.

35. Charles has a rectangular room that measures 10 feet by 11 feet. One seventh of the perimeter of the room is taken up by bookshelves. How many feet of bookshelves does Charles have in his room?

Multiply (fraction of a fraction).

1. $\dfrac{1}{4}$ of $\dfrac{1}{3}$ = ——

2. $\dfrac{3}{4} \times \dfrac{2}{5}$ = ——

3. $\dfrac{2}{7} \times \dfrac{1}{4}$ = ——

4. $\dfrac{2}{5}$ of $\dfrac{3}{7}$ = ——

5. $\dfrac{3}{5} \times \dfrac{1}{6}$ = ——

6. $\dfrac{2}{5} \times \dfrac{1}{3}$ = ——

Add or subtract.

7. $\dfrac{3}{11} + \dfrac{1}{4}$ = ——

8. $\dfrac{4}{5} - \dfrac{1}{6}$ = ——

9. $\dfrac{1}{7} + \dfrac{2}{3} + \dfrac{1}{2}$ = ——

Use the Rule of Four to make the denominators the same. Then compare the fractions.

10. $\dfrac{2}{3}$ ◯ $\dfrac{5}{8}$

11. $\dfrac{3}{4}$ ◯ $\dfrac{9}{12}$

12. $\dfrac{5}{6}$ ◯ $\dfrac{7}{9}$

Estimate and then multiply to find the exact answer.

13.   6 1 2 →
      × 5 4

14.   1 2 4 →
      × 3 6

15.   9 5 7 →
      × 1 3

16. One third of the customers at the ice cream store bought vanilla, and two fifths of them bought chocolate. What part of the customers bought vanilla or chocolate ice cream?

17. The sides of a triangle measure 10 feet, 12 feet, and 12 feet. What is the perimeter of the triangle?

18. Three eighths of the guests at the picnic ate hamburgers. One half of those people had mustard on their hamburgers. What part of the people at the picnic had hamburgers with mustard?

19. If there were 48 people at the picnic (#18), how many had hamburgers? How many had hamburgers with mustard?

20. Each of the 48 people at the picnic contributed $15 for food and other expenses. How much money was collected?

Divide. If the denominators are different, use the Rule of Four first to make them the same.

1. $\dfrac{2}{5} \div \dfrac{1}{5} =$ _____

2. $\dfrac{5}{8} \div \dfrac{2}{3} =$ _____

3. $\dfrac{1}{2} \div \dfrac{1}{4} =$ _____

4. $\dfrac{1}{4} \div \dfrac{1}{3} =$ _____

5. $\dfrac{3}{5} \div \dfrac{2}{9} =$ _____

6. $\dfrac{4}{5} \div \dfrac{2}{3} =$ _____

Multiply.

7. $\dfrac{3}{4} \times \dfrac{1}{4} =$ _____

8. $\dfrac{2}{3} \times \dfrac{1}{5} =$ _____

9. $\dfrac{2}{9} \times \dfrac{1}{2} =$ _____

Add or subtract.

10. $\dfrac{4}{7} + \dfrac{3}{8} =$ _____

11. $\dfrac{5}{9} - \dfrac{1}{4} =$ _____

12. $\dfrac{2}{5} + \dfrac{3}{10} + \dfrac{1}{2} =$ _____

Round and estimate. Do not worry about any remainders.

13. $38\overline{)541}$ →

14. $26\overline{)876}$ →

15. $55\overline{)605}$ →

16. If a board is $\frac{1}{12}$ of a foot thick, how many pieces are in a pile $\frac{1}{2}$ foot high?

17. Brooke has a piece of ribbon that is $\frac{7}{8}$ of a yard long. How many $\frac{1}{8}$-yard-long pieces can she cut from the piece of ribbon?

18. Jillian walked $\frac{3}{10}$ of a mile and ran $\frac{3}{5}$ of a mile. Then she roller-skated $\frac{1}{10}$ of a mile. How far did Jillian travel in all?

19. The first time Ned checked, he saw that $\frac{7}{8}$ of the pizza was left in the pan. Later he noticed that there was only $\frac{1}{4}$ of the pizza left in the pan. What part of a pizza was taken since he checked the first time?

20. There are 24 hours in a day. How many hours are in 365 days?

LESSON TEST

Use the divisibility rules to answer the questions. Write yes or no in the blanks.

1. Is 17 divisible by 2? ____

2. Is 105 divisible by 5? ____

3. Is 23 divisible by 3? ____

4. Is 288 divisible by 9? ____

List all the factors for each number and then underline the common factors. Write the greatest common factor (GCF) for each pair of numbers in the blank.

5. 8:

   24:

   The GCF of 8 and 24 is ____ .

6. 10:

   20:

   The GCF of 10 and 20 is ____ .

7. 39:

   15:

   The GCF of 39 and 15 is ____ .

8. 14:

   28:

   The GCF of 14 and 28 is ____ .

Multiply.

9. $\dfrac{1}{8} \times \dfrac{2}{7} =$ _____

10. $\dfrac{2}{9} \times \dfrac{4}{5} =$ _____

11. $\dfrac{3}{6} \times \dfrac{1}{6} =$ _____

Divide using the Rule of Four.

12. $\dfrac{1}{8} \div \dfrac{1}{2} =$ _____

13. $\dfrac{2}{3} \div \dfrac{1}{6} =$ _____

14. $\dfrac{5}{6} \div \dfrac{3}{10} =$ _____

Use the Rule of Four to make the denominators the same and then compare the fractions.

15. $\dfrac{3}{4} \bigcirc \dfrac{9}{12}$

16. $\dfrac{3}{8} \bigcirc \dfrac{4}{10}$

17. $\dfrac{1}{2} \bigcirc \dfrac{5}{9}$

Divide. Include a fraction in the answer if the problem does not divide evenly.

18. $26\overline{)673}$

19. $82\overline{)390}$

20. $51\overline{)768}$

Find the GCF of the numerator and denominator and use it to simplify each fraction.

1. $\dfrac{8\div}{10\div}$ ———— = ——

2. $\dfrac{12\div}{20\div}$ ———— = ——

3. $\dfrac{27\div}{30\div}$ ———— = ——

4. $\dfrac{20\div}{30\div}$ ———— = ——

5. $\dfrac{15\div}{25\div}$ ———— = ——

6. $\dfrac{36\div}{48\div}$ ———— = ——

Use the divisibility rules to answer the questions. Write *yes* or *no* in the blanks.

7. Is 18 divisible by 3? ____

8. Is 556 divisible by 5? ____

9. Is 41 divisible by 2? ____

10. Is 372 divisible by 9? ____

Solve and then use the GCF to simplify your answers.

11. $\dfrac{3}{6} + \dfrac{1}{4} =$ ———— = ——

12. $\dfrac{2}{9} + \dfrac{1}{6} =$ ———— = ——

13. $\dfrac{2}{3} \times \dfrac{5}{6} =$ —— $=$ ——

14. $\dfrac{5}{8} \times \dfrac{1}{10} =$ —— $=$ ——

15. $\dfrac{1}{5} \div \dfrac{8}{10} =$ —— $=$ ——

16. $\dfrac{4}{7} \div \dfrac{2}{3} =$ —— $=$ ——

Multiply. You may use lined paper if you wish.

17. $\begin{array}{r} 573 \\ \times\ 612 \\ \hline \end{array}$

18. $\begin{array}{r} 3{,}682 \\ \times\ 694 \\ \hline \end{array}$

19. Three eighths of the wall was left to be painted. Arik painted one half of what was left to be finished. What part of the whole wall did Arik paint?

20. Colleen walked $\dfrac{6}{8}$ of a mile. She divided her walk into $\dfrac{1}{4}$ mile sections so she could stop to stretch. Into how many $\dfrac{1}{4}$ mile sections did she divide her walk?

Find the prime factors using whatever method you wish.

1. 50 = _____

2. 48 = _____

3. 63 = _____

4. 99 = _____

Simplify the fractions using prime factors.

5. $\dfrac{20}{30}$ = _____ = _____

6. $\dfrac{63}{81}$ = _____ = _____ = _____

7. $\dfrac{36}{42}$ = _____ = _____ = _____

Solve and then simplify using the GCF or prime factors.

8. $\dfrac{1}{6} + \dfrac{5}{12}$ = _____ = _____

9. $\dfrac{5}{7} \div \dfrac{5}{6}$ = _____ = _____

10. $\dfrac{3}{4} \times \dfrac{1}{3}$ = _____ = _____

Use the Rule of Four to make the denominators the same and then compare the fractions.

11. $\dfrac{4}{7}$ ◯ $\dfrac{3}{8}$

12. $\dfrac{5}{10}$ ◯ $\dfrac{6}{11}$

13. $\dfrac{1}{4}$ ◯ $\dfrac{2}{15}$

Divide. Include a fraction in the answer if the problem does not divide evenly. You may use lined paper if you wish.

14. $8\,3\,\overline{\smash{)}\,5{,}3\,1\,1}$

15. $3\,6\,\overline{\smash{)}\,8{,}8\,5\,6}$

16. $5\,1\,2\,\overline{\smash{)}\,6{,}5\,3\,2}$

17. Paul planted 25 strawberry plants. If each plant yielded 17 quarts of berries this summer, how many quarts of berries did Paul get in all?

18. Becky drove for $\frac{1}{7}$ of her trip and flew for $\frac{7}{10}$ of it. She walked for the rest of the trip. For what part of the trip was Becky walking?

19. One half of the customers in the store were shopping for school supplies. One third of those customers bought backpacks. What part of the total customers bought backpacks?

20. Two sevenths of the total customers in #19 bought dictionaries. Which group was greater, those who bought backpacks or those who bought dictionaries?

Write the length of each line beneath the ruler. Simplify if possible. If you cannot simplify, leave the second blank empty.

1.

____ in = ____ in

2.

____ in = ____ in

3.

____ in = ____ in

4.

____ in = ____ in

5.   0"   1"

____ in = ____ in

6.   0"   1"

____ in = ____ in

Find the prime factors using whatever method you wish.

7. 24 = _____

8. 76 = _____

9. 48 = _____

Simplify the fractions using prime factors.

10. $\dfrac{32}{54}$ = _____ = _____ = _____

Find the area of each rectangle.

11. ┌─────────┐ 15"   A = _____
    └─────────┘
        32"

12. ┌─────────┐ 4'   A = _____
    └─────────┘
       10'

13. ┌─────────┐ 24 yd   A = _____
    └─────────┘
       49 yd

14. Is 168 divisible by nine?

15. What is the GCF of 16 and 44?

16. Two fifths of the days in April were cloudy. It rained on $\frac{3}{4}$ of the cloudy days. What part of the days in April was rainy?

17. A grocer bought 650 pounds of potatoes from a local farmer. If he put them in 10-pound bags for sale, how many bags did he use?

18. Wade wants to put a fence around his garden. Does he need to find the area or the perimeter of the garden?

Convert each mixed number to an improper fraction.

1. $1\dfrac{2}{9} = \dfrac{}{9} + \dfrac{}{9} =$ ——

2. $3\dfrac{3}{5} = \dfrac{}{5} + \dfrac{}{5} + \dfrac{}{5} + \dfrac{}{5} =$ ——

3. $1\dfrac{1}{3} = \dfrac{}{3} + \dfrac{}{3} =$ ——

4. $2\dfrac{1}{4} = \dfrac{}{4} + \dfrac{}{4} + \dfrac{}{4} =$ ——

Convert each improper fraction to a mixed number.

5. $\dfrac{7}{4} = \dfrac{}{4} + \dfrac{}{4} =$ ——

6. $\dfrac{9}{2} = \dfrac{}{2} + \dfrac{}{2} + \dfrac{}{2} + \dfrac{}{2} + \dfrac{}{2} =$ ——

7. $\dfrac{13}{5} = \dfrac{}{5} + \dfrac{}{5} + \dfrac{}{5} =$ ——

8. $\dfrac{23}{6} = \dfrac{}{6} + \dfrac{}{6} + \dfrac{}{6} + \dfrac{}{6} =$ ——

Write the length of each line beneath the ruler. Simplify if possible. If you cannot simplify, leave the second blank empty.

9.

_____ in = _____ in

10.

_____ in = _____ in

Simplify the fractions using any method you wish.

11. $\dfrac{3}{9} =$ ——

12. $\dfrac{20}{28} =$ ——

13. $\dfrac{18}{32} =$ ——

Find the area of a square with the given side length.

14. S = 16 inches   A = _____

15. S = 3 miles   A = _____

Solve.

16. 8^2 = _____

17. 6^2 = _____

18. Is 213 divisible by three?

19. What is the GCF of 15 and 30?

20. What are the prime factors of 45?

Write the length of the line.

1. _____ in

2. _____ in

Convert each mixed number to an improper fraction.

3. $1\dfrac{1}{8} = \dfrac{}{8} + \dfrac{}{8} =$ _____

4. $2\dfrac{5}{6} = \dfrac{}{6} + \dfrac{}{6} + \dfrac{}{6} =$ _____

Convert each improper fraction to a mixed number.

5. $\dfrac{9}{4} = \dfrac{}{4} + \dfrac{}{4} + \dfrac{}{4} =$ _____

6. $\dfrac{24}{7} = \dfrac{}{7} + \dfrac{}{7} + \dfrac{}{7} + \dfrac{}{7} =$ _____

Solve. Simplify if possible and write any improper fractions as mixed numbers.

7. $\dfrac{3}{7} + \dfrac{2}{9} =$ _____

8. $\dfrac{1}{8} + \dfrac{1}{6} =$ _____

9. $\dfrac{1}{3} - \dfrac{2}{8} =$ _____

10. $\dfrac{5}{6} \times \dfrac{11}{12} =$ _____

11. $\dfrac{1}{2} \div \dfrac{3}{10} =$ _____

12. $\dfrac{2}{3} \times \dfrac{7}{20} =$ _____

Find the area and perimeter of each figure.

4'
10'

13. A = _____

14. P = _____

23"
23"

15. A = _____

16. P = _____

13'
5'
12'

17. A = _____

18. P = _____

19. What number is 16^2 equal to?

20. What is the GCF of 36 and 42?

Divide. Simplify answers when possible and write any improper fractions as mixed numbers.

1. $\dfrac{1}{2} \div \dfrac{5}{9} =$ _____

2. $\dfrac{3}{7} \div \dfrac{5}{6} =$ _____

3. $\dfrac{4}{7} \div \dfrac{5}{14} =$ _____

4. $\dfrac{7}{9} \div \dfrac{2}{3} =$ _____

5. $\dfrac{3}{4} \div \dfrac{1}{8} =$ _____

6. $\dfrac{1}{4} \div \dfrac{3}{5} =$ _____

Multiply. Simplify answers when possible.

7. $\dfrac{1}{4} \times \dfrac{1}{3} =$

8. $\dfrac{4}{5} \times \dfrac{7}{10} =$

9. $\dfrac{2}{5} \times \dfrac{1}{2} =$

10. $\dfrac{3}{4} \times \dfrac{1}{4} =$

11. $\dfrac{3}{7} \times \dfrac{2}{9} =$

12. $\dfrac{1}{8} \times \dfrac{1}{6} =$

Use the divisibility rules to answer the questions. Write *yes* or *no* in the blanks.

13. Is 18 divisible by 2? ____

14. Is 552 divisible by 5? ____

15. Is 102 divisible by 3? ____

16. Is 962 divisible by 9? ____

17. What is the GCF of 28 and 54?

18. What are the prime factors of 60?

Simplify the fractions using any method you wish.

19. $\dfrac{4}{12}$ = ——

20. $\dfrac{15}{20}$ = ——

21. $\dfrac{18}{42}$ = ——

Convert each mixed number to an improper fraction.

22. $1\dfrac{3}{4}$ = $\dfrac{}{4}$ + $\dfrac{}{4}$ = ——

23. $2\dfrac{7}{9}$ = $\dfrac{}{9}$ + $\dfrac{}{9}$ + $\dfrac{}{9}$ = ——

Convert each improper fraction to a mixed number.

24. $\dfrac{13}{5}$ = $\dfrac{}{5}$ + $\dfrac{}{5}$ + $\dfrac{}{5}$ = ——

25. $\dfrac{10}{3}$ = $\dfrac{}{3}$ + $\dfrac{}{3}$ + $\dfrac{}{3}$ + $\dfrac{}{3}$ = ——

Write the length of the line.

26. ____ in

27. ____ in

28. ____ in

Find the area of each figure.

29. 3'

7'

A = _____

30. 12"

12"

A = _____

31. 6' 10'

8'

A = _____

32. What number is 9^2 equal to?

33. Marcy had $\frac{3}{4}$ of her birthday cake left over. She wanted to give each of her guests $\frac{1}{16}$ of a whole cake. How many people can she serve?

34. Before Marcy could serve her guests, her dog came along and ate one half of the cake that was left over (#33). What part of a whole cake did the dog eat?

Add or subtract. Continue to estimate your answers mentally.

1.
$$
\begin{array}{r}
2\frac{1}{10} \\
+\,6\frac{6}{10} \\
\hline
\end{array}
$$

2.
$$
\begin{array}{r}
4\frac{3}{5} \\
-\,3\frac{1}{5} \\
\hline
\end{array}
$$

3.
$$
\begin{array}{r}
5\frac{2}{7} \\
+\,1\frac{4}{7} \\
\hline
\end{array}
$$

Convert each mixed number to an improper fraction. Write out the steps only if you need to.

4. $6\frac{1}{2}$ = ——

5. $5\frac{2}{3}$ = ——

6. $1\frac{9}{10}$ = ——

Convert each improper fraction to a mixed number.

7. $\frac{21}{4}$ = ——

8. $\frac{11}{5}$ = ——

9. $\frac{15}{8}$ = ——

Solve. Simplify if possible.

10. $\dfrac{5}{6} \times \dfrac{1}{3} =$ _____

11. $\dfrac{1}{10} \div \dfrac{3}{10} =$ _____

12. $\dfrac{8}{9} \times \dfrac{5}{6} =$ _____

Find the volume of each cube.

13.   5" 5" 5"   V = _____

14.   8' 8' 8'   V = _____

15.   13' 13' 13'   V = _____

16. Andrew built a vacation house in the shape of a cube measuring 17 feet by 17 feet by 17 feet. What was the volume of his new house?

17. What are the prime factors of 95?

18. Briana went shopping and bought $4\frac{3}{10}$ pounds of hamburger and $2\frac{6}{10}$ pounds of steak. How many pounds of meat did she buy?

19. A square is $\frac{3}{4}$ of a mile on each side. What is the area of the square?

20. One hundred twenty people came to watch the ball game at the local park. Five eighths of them were parents of the players. How many people were parents of the players?

Add the mixed numbers.

1. $6\frac{4}{7}$

$+7\frac{5}{7}$

2. $13\frac{4}{9}$

$+5\frac{8}{9}$

3. $21\frac{11}{12}$

$+3\frac{1}{12}$

4. $4\frac{3}{5}$

$+8\frac{4}{5}$

5. $9\frac{2}{3}$

$+8\frac{2}{3}$

6. $6\frac{7}{8}$

$+7\frac{3}{8}$

Solve.

7. $\frac{3}{4} - \frac{1}{5} =$ _____

8. $\frac{1}{10} + \frac{1}{5} =$ _____

9. $\frac{7}{9} - \frac{1}{2} =$ _____

10. $\frac{3}{10} \div \frac{1}{10} =$ _____

11. $\dfrac{5}{6} \times \dfrac{1}{3} = $ ____

12. $\dfrac{3}{4} \div \dfrac{1}{3} = $ ____

Solve. Remember that the multiplication sign means the same as "of."

13. $\dfrac{3}{8} \times 64 = $ ____

14. $\dfrac{1}{6} \times 42 = $ ____

15. $\dfrac{7}{10} \times 80 = $ ____

16. Kara built a rectangular fish pond in her garden. It was eight feet long, five feet wide, and two feet deep. How many cubic feet of water would be required to fill it to the top?

17. At 63 pounds per cubic foot, how much will the water in #16 weigh?

18. What is the GCF of 42 and 56?

19. What are the prime factors of 18?

20. Aaron ran $2\frac{4}{5}$ of a mile. Tony ran as far as Aaron, and then he ran $1\frac{3}{5}$ miles farther. How many miles did Tony run?

Add or subtract the mixed numbers and regroup as needed.

1. $2\dfrac{1}{8}$

    $-\,1\dfrac{5}{8}$
    _____

2. $5\dfrac{2}{7}$

    $-\ \ \dfrac{3}{7}$
    _____

3. 3

    $-\,2\dfrac{1}{4}$
    _____

4. $2\dfrac{5}{8}$

    $+\,2\dfrac{4}{8}$
    _____

5. $7\dfrac{3}{5}$

    $+\,4\dfrac{4}{5}$
    _____

6. $8\dfrac{3}{6}$

    $+\,9\dfrac{3}{6}$
    _____

Solve.

7. $\dfrac{4}{5} \div \dfrac{2}{3} = $ ____

8. $\dfrac{2}{11} \times \dfrac{1}{5} = $ ____

9. $\dfrac{1}{2} + \dfrac{2}{3} + \dfrac{5}{6} = $ ____

Use the Rule of Four to make the denominators the same. Then compare the fractions.

10. $\dfrac{1}{4}$ $\bigcirc$ $\dfrac{2}{5}$

11. $\dfrac{3}{7}$ $\bigcirc$ $\dfrac{1}{8}$

12. $\dfrac{5}{12}$ $\bigcirc$ $\dfrac{4}{10}$

Fill in the blanks.

13. 99 ft = ____ yd

14. 26 yd = ____ ft

15. 9 yd = ____ ft

16. A rectangle is $\frac{2}{3}$ of a mile long and $\frac{1}{3}$ of a mile wide. What is the area of the rectangle?

17. What is the perimeter of the rectangle in #16?

18. A restaurant sold $9\frac{5}{8}$ pies on Monday and $6\frac{4}{8}$ pies on Tuesday. How many pies were sold during those two days?

19. Melissa bought $15\frac{1}{8}$ yards of fabric for curtains. If she uses only $11\frac{3}{8}$ yards, how much fabric will she have left over?

20. If Mark is six feet tall, how many yards tall is he?

Subtract using the "same difference theorem."

1. $8 \dfrac{5}{8}$ +   =

    $- \ 3 \dfrac{7}{8}$ +   =
    _____

2. 14 +   =

    $- \ 10 \dfrac{1}{3}$ +   =
    _____

Add or subtract. Use whichever method you prefer to subtract.

3. $2\dfrac{3}{4}$

    $+ \ 3\dfrac{3}{4}$
    _____

4. $7\dfrac{1}{3}$

    $- \ 3\dfrac{2}{3}$
    _____

5. $4\dfrac{3}{5}$

    $+ \ 1\dfrac{4}{5}$
    _____

Add or subtract.

6. $\dfrac{2}{3} + \dfrac{3}{4} =$ _____

7. $\dfrac{3}{4} + \dfrac{5}{6} =$ _____

8. $\dfrac{1}{2} + \dfrac{4}{7} =$ _____

9. $\dfrac{7}{8} - \dfrac{1}{3} =$ _____

10. $\dfrac{5}{12} - \dfrac{2}{10} =$ _____

11. $\dfrac{1}{6} - \dfrac{1}{7} =$ _____

Fill in the blanks.

12. 32 qt = ____ pt

13. 34 pt = ____ qt

14. 7 yd = ____ ft

15. A farmer had $9\frac{3}{4}$ tons of hay the first time he cut the field and $6\frac{3}{4}$ tons the second time he cut it. What was the total amount of hay he harvested?

16. Luke had $4\frac{2}{5}$ gallons of gasoline in his tank when he started his trip. If he used $1\frac{4}{5}$ gallons, how much was left in the tank after the trip?

17. Is 616 divisible by 3?

18. A cube measures nine feet on a side. Give the volume in cubic yards. (Convert feet to yards first.)

19. A square measures $2\frac{1}{4}$ feet on each side. What is the perimeter of the square?

20. What number does 15^2 equal?

LESSON TEST

21

Add the mixed numbers.

1. $9\dfrac{1}{3}$

    $+\ 6\dfrac{1}{4}$

2. $4\dfrac{2}{3}$

    $+\ 1\dfrac{3}{5}$

3. $9\dfrac{3}{5}$

    $+\ 2\dfrac{7}{10}$

4. $12\dfrac{5}{11}$

    $+\ 4\dfrac{5}{8}$

Subtract. Use whichever method you prefer.

5. $5\dfrac{1}{5}$

    $-\ 2\dfrac{3}{5}$

6. $15\dfrac{7}{9}$

    $-\ 6\dfrac{2}{9}$

7. 7

    $-\ 3\dfrac{1}{5}$

Divide or multiply.

8. $\dfrac{1}{2} \div \dfrac{1}{4} =$ _____

9. $\dfrac{8}{12} \div \dfrac{1}{3} =$ _____

10. $\dfrac{3}{5} \div \dfrac{1}{10} =$ _____

11. $\dfrac{4}{6} \times \dfrac{3}{5} =$ _____

12. $\dfrac{1}{2} \times \dfrac{1}{3} =$ _____

13. $\dfrac{3}{4} \times \dfrac{1}{6} =$ _____

Fill in the blanks.

14. 81 ft = _____ yd

15. 54 pt = _____ qt

16. 18 qt = _____ pt

17. How many yards are there in 92 feet? Include a fraction in the answer if the problem does not divide evenly.

18. Duncan drove $20\frac{1}{2}$ miles and then walked $2\frac{3}{4}$ miles. How far did he travel in all?

19. Each side of a square is $\frac{2}{3}$ of a foot long. Find the area and the perimeter of the square.

20. Mental Math! 3 times 4, times 2, plus 1, divided by 5, equals _____ .

Solve.

1. $4\frac{1}{4}$
   $-2\frac{3}{4}$

2. $4\frac{1}{2}$
   $-1\frac{1}{3}$

3. $8\frac{5}{9}$
   $-3\frac{2}{9}$

4. $16\frac{3}{10}$
   $-5\frac{2}{5}$

5. $1\frac{1}{2}$
   $+3\frac{2}{3}$

6. $4\frac{2}{3}$
   $+3\frac{4}{5}$

7. $6\frac{3}{4}$
   $+9\frac{7}{8}$

Solve.

8. $\dfrac{5}{16} \div \dfrac{3}{4} =$ _____

9. $\dfrac{1}{2} \times \dfrac{2}{3} =$ _____

10. $\dfrac{7}{8} \div \dfrac{5}{12} =$ _____

Convert each mixed number to an improper fraction.

11. $9\dfrac{5}{8} =$ ——

12. $25\dfrac{2}{3} =$ ——

13. $10\dfrac{3}{4} =$ ——

Fill in the blanks. Include a fraction in the answer if the problem does not divide evenly.

14. 14 qt = _____ gal

15. 8 gal = _____ qt

16. 20 pt = _____ qt

17. The area of a rectangle is 672 square yards. The width is 16 yards. What is the length of the other side of the rectangle?

18. Give the width of the rectangle in #17 in feet.

19. What is the GCF of 14 and 56?

20. Mental Math! 9 plus 1, minus 2, times 7, plus 4, equals _____ .

LESSON TEST

Divide using the reciprocal. Then check your work by using the Rule of Four.

Reciprocal

Rule of Four

1. $\dfrac{3}{4} \div \dfrac{1}{2} =$

2. $\dfrac{3}{4} \div \dfrac{1}{2} =$

3. $\dfrac{4}{5} \div \dfrac{2}{3} =$

4. $\dfrac{4}{5} \div \dfrac{2}{3} =$

Convert mixed numbers to improper fractions and divide using the reciprocal.

5. $1\dfrac{2}{3} \div \dfrac{5}{11} =$

6. $1\dfrac{7}{8} \div \dfrac{1}{8} =$

7. $2\dfrac{3}{4} \div \dfrac{1}{2} =$

8. $6\dfrac{1}{5} \div 3\dfrac{3}{8} =$

Solve.

9. $\quad 20\dfrac{1}{5}$
$-10\dfrac{3}{4}$
_____

10. $\quad 9\dfrac{1}{8}$
$-4\dfrac{1}{3}$
_____

11. $5\dfrac{3}{7}$
$+\ 5\dfrac{1}{4}$
_____

Fill in the blanks. Include a fraction in the answer if the problem does not divide evenly.

12. 17 oz = ____ lb

13. 28 pt = ____ qt

14. 10 yd = ____ ft

15. 5 lb = ____ oz

16. 9 qt = ____ pt

17. 19 qt = ____ gal

18. Bria has $3\dfrac{1}{8}$ pounds of chocolates. If she divides them into portions that each weigh $\dfrac{5}{8}$ of a pound, how many people can she treat?

19. Oliver's rectangular garden cart measures four feet by two feet by one foot. If he fills it to the top, how many cubic feet of mulch can he haul in one load?

20. Mental Math! 6 times 7, plus 3, divided by 9, plus 2, equals _____ .

Solve.

1. $3\dfrac{3}{8}$
    $+\ 4\dfrac{1}{8}$

2. $7\dfrac{3}{5}$
    $+\ 4\dfrac{4}{5}$

3. $15\dfrac{2}{7}$
    $+\ 11\dfrac{1}{2}$

4. $8\dfrac{5}{6}$
    $+\ 5\dfrac{2}{3}$

Solve.

5. $9\dfrac{7}{8}$
    $-\ 3\dfrac{5}{8}$

6. 3
    $-\ 1\dfrac{2}{5}$

7. $35\dfrac{1}{4}$
    $-\ 21\dfrac{2}{3}$

8. $7\dfrac{1}{6}$
    $-\ 6\dfrac{3}{7}$

Fill in the blanks. Include a fraction in the answer if the problem does not divide evenly.

9. 48 oz = ____ lb

10. 23 pt = ____ qt

11. 9 yd = ____ ft

12. 4 lb = ____ oz

13. 16 qt = ____ pt

14. 20 qt = ____ gal

Divide using the reciprocal and then check your work by using the Rule of Four.

Reciprocal

Rule of Four

15. $\dfrac{3}{6} \div \dfrac{1}{2} =$

16. $\dfrac{3}{6} \div \dfrac{1}{2} =$

17. $\dfrac{5}{8} \div \dfrac{2}{3} =$

18. $\dfrac{5}{8} \div \dfrac{2}{3} =$

Convert any mixed numbers to improper fractions and divide using the reciprocal.

19. $6\dfrac{1}{3} \div 2\dfrac{3}{8} =$

20. $2\dfrac{3}{5} \div 1\dfrac{9}{10} =$

21. $5\dfrac{5}{6} \div \dfrac{5}{6} =$

22. $3\dfrac{1}{2} \div 1\dfrac{3}{7} =$

23. What is the reciprocal of 8?

24. A pool is 15 feet long, 10 feet wide, and 6 feet deep. How many cubic feet of water are required to fill it to the brim?

25. Joseph has $4\frac{1}{2}$ pizzas for his party. If there are eight people to serve, how much pizza can each one have?

26. Christopher jogged $2\frac{1}{3}$ miles and ran $1\frac{5}{6}$ miles. How far did he run altogether?

27. At the beginning of the day, Ethan had $5\frac{1}{2}$ rows to weed in his garden. If he has finished $2\frac{2}{3}$ rows, how much is left to do?

28. Mental Math! 8 times 6, minus 4, divided by 2, plus 3, equals _____ .

Use the multiplicative inverse to solve each equation. Check your work by replacing the unknown with the solution.

1. $10J = 200$

2. Check for #1

3. $6U = 24$

4. Check for #3

5. $5K = 95$

6. Check for #5

Divide using the reciprocal. Then check your work by using the Rule of Four.

Reciprocal

Rule of Four

7. $\dfrac{1}{3} \div \dfrac{2}{5} =$

8. $\dfrac{1}{3} \div \dfrac{2}{5} =$

Convert any mixed numbers to improper fractions and divide using the reciprocal.

9. $1\dfrac{1}{4} \div 3\dfrac{1}{2} =$

10. $4\dfrac{1}{6} \div 4\dfrac{4}{5} =$

Solve.

11. $3\dfrac{1}{3}$
    $-1\dfrac{2}{3}$
    _____

12. $8\dfrac{2}{3}$
    $+5\dfrac{3}{4}$
    _____

13. $4\dfrac{1}{6}$
    $-1\dfrac{1}{2}$
    _____

Solve.

14. $\dfrac{2}{3} \times \dfrac{3}{5} =$

15. $\dfrac{2}{7} \times \dfrac{3}{4} =$

16. $\dfrac{5}{8} \times \dfrac{1}{3} =$

17. If Richard is six feet tall, how many inches tall is he?

18. Three times Jacki's age is 120. How old is Jacki? Write an equation and solve.

19. How many inches are there in one half of a foot?

20. A square measures two thirds of a mile on each side. Give the area and the perimeter of the square.

Convert any mixed numbers to improper fractions and multiply. Use canceling as a short cut.

1. $\dfrac{2}{3} \times \dfrac{3}{4} \times \dfrac{2}{7} =$

2. $1\dfrac{5}{7} \times \dfrac{1}{5} \times 1\dfrac{3}{4} =$

3. $\dfrac{1}{8} \times 1\dfrac{1}{7} \times \dfrac{4}{7} =$

4. $1\dfrac{1}{5} \times \dfrac{5}{6} \times 1\dfrac{1}{2} =$

Use the multiplicative inverse to solve each equation. Check your work.

5. $7L = 77$

6. Check for #5

7. $13Y = 195$

8. Check for #7

Divide using whichever method you prefer.

9. $1\dfrac{9}{16} \div \dfrac{5}{8} =$

10. $6\dfrac{1}{3} \div 2\dfrac{3}{8} =$

Solve.

11. $6\frac{5}{8}$

   $-3\frac{7}{8}$

12. $12\frac{1}{4}$

   $-11\frac{3}{4}$

13. $1\frac{1}{3}$

   $+1\frac{2}{3}$

Fill in the blanks. Include a fraction in the answer if the problem does not divide evenly.

14. 2 tons = _____ lb

15. 24 in = _____ ft

16. $\frac{1}{4}$ ton = _____ lb

17. Lillian's car weighs $1\frac{1}{2}$ tons. How many pounds does it weigh?

18. Five eighths of the money was for food. One tenth of that was used for eating out, and Rose got four fifths of that amount for herself because she traveled a lot. What part of the total money did Rose get for eating out?

19. If $6\frac{1}{2}$ dozen doughnuts are divided among 13 people, how many doughnuts does each person receive?

20. Mental Math! 28 divided by 4, plus 1, times 8, minus 4, equals _____ .

Use the multiplicative inverse to solve each equation. Check your work.

1. 6B + 10 = 40

2. Check for #1

3. 8N − 13 = 3

4. Check for #3

5. 6C + 8 = 50

6. Check for #5

Convert any mixed numbers to improper fractions and multiply. Use canceling as a short cut.

7. $1\frac{1}{5} \times 3\frac{1}{3} \times \frac{3}{11} =$

8. $\frac{1}{2} \times \frac{5}{12} \times 3\frac{3}{7} =$

Divide using the method you prefer.

9. $\frac{1}{5} \div \frac{8}{25} =$

10. $6\frac{2}{3} \div 2\frac{6}{7} =$

Fill in the blanks. Include a fraction in the answer if the problem does not divide evenly.

11. 3 mi = _____ ft

12. 7 lb = _____ oz

13. $3\frac{3}{4}$ ft = _____ in

14. A farmer's truck was carrying $2\frac{1}{4}$ tons of hay and $3\frac{1}{4}$ tons of straw. How many tons were loaded on the truck?

15. Is 349 divisible by 9?

16. What are the prime factors of 48?

17. What is the GCF of 18 and 24?

18. A square measures $1\frac{1}{2}$ miles on each side. What is the area of the square?

19. What number does 16^2 equal?

20. Mental Math! 56 divided by 7, plus 10, minus 3, plus 15, equals _____

Find the area and circumference of each circle. Use $\frac{22}{7}$ as an approximate value of π.

1. Area ≈ _____

2. Circumference ≈ _____

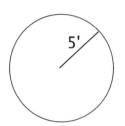

3. Area ≈ _____

4. Circumference ≈ _____

5. Area ≈ _____

6. Circumference ≈ _____

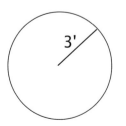

7. Area ≈ _____

8. Circumference ≈ _____

Use the multiplicative inverse to solve the equation. Check your work.

9. $5R - 17 = 38$

10. Check for #9

Solve.

11. $1\dfrac{1}{3} \times 1\dfrac{1}{2} =$

12. $1\dfrac{2}{3} \times 2\dfrac{1}{3} \times \dfrac{9}{10} =$

Solve.

13. $\dfrac{7}{10} \div \dfrac{7}{12} =$

14. $4\dfrac{2}{5} \div \dfrac{7}{9} =$

Use <, >, or = to compare the measures.

15. 19 qt $\bigcirc$ 5 gal

16. 3 tons $\bigcirc$ 5,000 lb

17. $\dfrac{2}{3}$ mi $\bigcirc$ 3,720 ft

18. Keith drove $8\dfrac{2}{3}$ miles from home and then started to walk back. After he had walked $6\dfrac{1}{2}$ miles, how much farther did he have to travel to get home?

19. How many feet did Keith have left to travel (#18)?

20. The base of a triangle measures 4 inches, and the height is $2\dfrac{1}{2}$ inches. What is the area of the triangle?

Use multiplication, addition, and subtraction as needed to solve each equation. Check your work.

1. $\dfrac{4}{15} W + 7 = 19$

2. Check for #1

3. $\dfrac{2}{3} R - 5 = 5$

4. Check for #3

5. $\dfrac{1}{2} X = 4$

6. Check for #5

Convert any mixed numbers to improper fractions and multiply.

7. $1\dfrac{7}{8} \times 2\dfrac{2}{3} =$

8. $\dfrac{4}{5} \times 3\dfrac{1}{3} \times 1\dfrac{7}{8} =$

Divide using the method you prefer.

9. $\dfrac{11}{18} \div \dfrac{3}{8} =$

10. $4\dfrac{1}{2} \div 2\dfrac{1}{2} =$

Use <, >, or = to compare the measures.

11. 3 lb $\bigcirc$ 32 oz

12. $\dfrac{1}{5}$ mi $\bigcirc$ 1,056 ft

13. $\dfrac{1}{4}$ gal $\bigcirc$ 2 qt

Find the average of each series of numbers.

14. 4, 2, 6, 8   Average = _____

15. 1, 3, 5, 7   Average = _____

16. 8, 10, 12   Average = _____

17. What is the approximate area of a circle with a radius of 28 yards?

18. What is the approximate circumference of the circle in #17?

19. What is the volume of a cube that measures $2\frac{3}{4}$ inches on each side?

20. Mental Math! 9 times 8, minus 2, divided by 10, times 4, equals _____ .

Convert each fraction to an equivalent fraction in hundredths and then to a percent.

1. $\dfrac{5}{5} = \dfrac{}{100} = \underline{\hphantom{000}}\%$

2. $\dfrac{3}{5} = \dfrac{}{100} = \underline{\hphantom{000}}\%$

Write each fraction in hundredths. Then write it as a decimal and as a percent.

3. $\dfrac{4}{5} = \dfrac{}{100} = \underline{\hphantom{000}} = \underline{\hphantom{000}}\%$

4. $\dfrac{3}{4} = \dfrac{}{100} = \underline{\hphantom{000}} = \underline{\hphantom{000}}\%$

Use multiplication, addition, and subtraction as needed to solve the equation. Check your work.

5. $\dfrac{3}{5}Y + 2 = 26$

6. Check for #5

Solve.

7. $\begin{array}{r} 7\frac{2}{5} \\ -\ 2\frac{4}{5} \\ \hline \end{array}$

8. $\begin{array}{r} 9 \\ -\ 6\frac{3}{4} \\ \hline \end{array}$

9. $\begin{array}{r} 10\frac{9}{10} \\ +\ 4\frac{3}{8} \\ \hline \end{array}$

Write the Arabic numeral that corresponds to each Roman numeral.

10. X _____

11. XIV _____

12. XXII _____

13. XXXV _____

Write the Roman numeral that corresponds to each Arabic numeral.

14. 3 _____

15. 17 _____

16. 25 _____

17. 19 _____

18. Sandra earned money for chores. This week she earned the following amounts on different days: 15 dollars, 10 dollars, 8 dollars, and 25 dollars. What is the average amount she earned per day?

19. What are the approximate area and circumference of a circle that has a radius of 14 inches?

20. The base of a triangle measures 16 feet, and the height is $5\frac{3}{4}$ feet. What is the area of the triangle?

Use multiplication, addition, and subtraction as needed to solve each equation. Check your work.

1. $\dfrac{4}{9}W + \dfrac{1}{3} = \dfrac{2}{3}$

2. Check for #1

3. $\dfrac{1}{2}K - \dfrac{3}{4} = 4\dfrac{1}{4}$

4. Check for #3

Write each fraction in hundredths. Then write it as a decimal and as a percent.

5. $\dfrac{1}{5} = \dfrac{}{100} =$ _____ = _____ %

6. $\dfrac{1}{4} = \dfrac{}{100} =$ _____ = _____ %

Solve.

7. $1\dfrac{7}{8} \div \dfrac{5}{8} =$ _____

8. $1\dfrac{2}{8} \times 3\dfrac{1}{5} \times 5\dfrac{1}{5} =$ _____

9. $\begin{array}{r} 8\dfrac{2}{3} \\ - 4\dfrac{5}{6} \\ \hline \end{array}$

10. $\begin{array}{r} 11\dfrac{1}{2} \\ + 6\dfrac{5}{9} \\ \hline \end{array}$

Write the Arabic numeral that corresponds to each Roman numeral.

11. LI _____

12. CXV _____

13. MDC _____

Write the Roman numeral that corresponds to each Arabic numeral.

14. 11 _____

15. 300 _____

16. 1,700 _____

17. How many feet are in two miles?

18. How many pounds are in three tons?

19. Mary's rectangular fish pond is 15 feet long, 10 feet wide, and $3\frac{1}{2}$ feet deep. What is the volume of the water if the pond is filled to the top?

20. Solve for the unknown: $8X - 8 = 40$

Solve for the unknown. Check your work.

1. $7L = 49$

2. Check for #1

3. $3R - 8 = 22$

4. Check for #3

5. $\dfrac{5}{8}X + 9 = 19$

6. Check for #5

7. $\dfrac{3}{10}Y - \dfrac{1}{5} = \dfrac{2}{5}$

8. Check for #7

9. $\dfrac{3}{4}T + \dfrac{3}{8} = 6\dfrac{3}{8}$

10. Check for #9

Solve.

11. $\dfrac{9}{10} \times \dfrac{5}{7} \times \dfrac{2}{3} = $ _____

12. $2\dfrac{1}{9} \times 2\dfrac{1}{4} \times 1\dfrac{1}{11} = $ _____

Write each fraction in hundredths. Then write it as a decimal and as a percent.

13. $\dfrac{3}{4} = \dfrac{}{100} = $ _____ $= $ _____ %

14. $\dfrac{1}{2} = \dfrac{}{100} = $ _____ $= $ _____ %

Find the area and circumference of the circle. Use $\frac{22}{7}$ as an approximate value of π.

15. Area ≈ _____

16. Circumference ≈ _____

Write the Arabic numeral that corresponds to each Roman numeral.

17. X _____

18. I _____

19. L _____

20. M _____

21. D _____

22. V _____

23. C _____

Answer the questions.

24. Ron walked for five miles. How many feet did he walk?

25. Farmer Brown bought $6\frac{1}{2}$ tons of cattle feed. How many pounds of feed is that?

26. Eric is $5\frac{3}{4}$ feet tall. How many inches tall is he?

27. A room is 169 inches wide. How many feet wide is the room?

28. Wesley got the following scores on his unit tests: 100, 90, 85, 97. What is his average score?

Solve.

1. $\dfrac{1}{2}$ of 24 = ____

2. $\dfrac{2}{3}$ of 18 = ____

3. $\dfrac{7}{8}$ of 64 = ____

Fill in the missing numbers in the numerators or denominators to make equivalent fractions.

4. $\dfrac{3}{4} = \dfrac{\quad}{\quad} = \dfrac{\quad}{\quad} = \dfrac{\quad}{16}$

5. $\dfrac{9}{10} = \dfrac{\quad}{\quad} = \dfrac{\quad}{\quad} = \dfrac{36}{\quad}$

Use the Rule of Four to compare the fractions. Write the correct symbol in the oval.

6. $\dfrac{5}{7} \bigcirc \dfrac{3}{5}$

7. $\dfrac{4}{8} \bigcirc \dfrac{3}{6}$

8. $\dfrac{4}{8} \bigcirc \dfrac{2}{3}$

Solve.

9. $\dfrac{3}{9} + \dfrac{5}{9} =$

10. $\dfrac{1}{2} + \dfrac{1}{4} + \dfrac{7}{8} =$

11. $\dfrac{4}{5} - \dfrac{1}{3} =$

12. $\dfrac{1}{3} \div \dfrac{1}{5} =$

13. $3\dfrac{1}{3} \div \dfrac{5}{18} =$

14. $3\dfrac{4}{5} \div 2\dfrac{7}{25} =$

Solve.

15. $\quad 7\dfrac{1}{4}$
    $-5\dfrac{3}{4}$
    _____

16. $\quad 9\dfrac{2}{3}$
    $+6\dfrac{5}{9}$
    _____

17. $\quad 5\dfrac{1}{5}$
    $-2\dfrac{5}{6}$
    _____

Write the length of the line.

18. _____ in

Solve for the unknown and check your work.

19. $7X + 9 = 44$

20. Check for #19

21. $\frac{3}{8}A - 8 = 13$

22. Check for #21

23. $\frac{5}{6}G + \frac{1}{6} = \frac{5}{12}$

24. Check for #23

Solve.

25. $\frac{5}{8} \times \frac{1}{3} \times \frac{3}{5} =$ _____

26. $\frac{4}{5} \times 2\frac{3}{4} \times 3\frac{1}{3} =$ _____

Write each fraction in hundredths. Then write it as a decimal and as a percent.

27. $\dfrac{4}{5} = \dfrac{}{100} = $ _____ $= $ _____ %   28. $\dfrac{1}{4} = \dfrac{}{100} = $ _____ $= $ _____ %

29. What is the GCF of 15 and 45?

30. What are the prime factors of 56?

31. Change $7\frac{2}{3}$ to an improper fraction.

32. Is 498 divisible by 9?

33. What is the approximate area of a circle with a radius of 21 feet?

34. What is the approximate circumference of a circle with a radius of 21 feet?